Desert Wildflowers

100 *Desert*

Photography & Text

SOUTHWESTERN

Wildflowers

in natural color

Natt N. Dodge

MONUMENTS   ASSOCIATION

Globe, Arizona

Library of Congress Catalog Card Number: 63-13471

Printed in the United States of America

Tyler Printing Co.   •   Phoenix, Arizona

Introduction

The Desert

When Webster defined a desert as a "dry, barren region, largely treeless and sandy" he was not thinking of the 50,000 square mile Great American Desert of the southwestern United States and northern Mexico. Most of it is usually dry and parts may be sandy, but as a whole, it is far from barren and treeless. Heavily vegetated with gray-green shrubs, small but robust trees, pygmy forests of grotesque cactuses and stiff-leaved yuccas, and myriads of herbaceous plants, the desert, following rainy periods, covers itself with a blanket of delicate, fragrant wildflowers. Edmund C. Jaegar, author of several books on deserts, reports that the California deserts alone support more than 700 species of flowering plants.

The late Dr. Forrest Shreve, for many years Director of the Desert Laboratory of the Carnegie Institution near Tucson, Arizona, defined a desert as "a region of deficient and uncertain rainfall." He divided the Great American Desert into four major sections: (1) *Chihuahuan* (chee-WAH-wahn), including the Mexican States of Chihuahua and Coahuila (coa-WHEE-lah), southwestern Texas, and south-central New Mexico; (2) *Sonoran,* including Baja California, southwestern Arizona, and northwestern Sonora; (3) *Mojave* (moh-HAH-vee), Colorado, including southeastern California and extreme southern Nevada; (4) *Great Basin,* including Nevada, Utah, southwestern Idaho and southeastern Oregon.

Since the steppes and mesas of the Great Basin Desert have generally lower temperatures, higher elevations, and greater precipitation than the other three sections, we are not including its flowers in this work.

Why and When Do Deserts Bloom?

The Great American Desert produces, when conditions are favorable, a gorgeous exhibition of wildflowers. Trees, shrubs, and herbs all contribute to the splendor of the display. Soil composition, slope and exposure, suitable temperatures, and adequate moisture are essential to plant growth and flower production.

Moisture is the uncertain factor, and years may pass without enough rainfall to stimulate plant growth. Rain of less than 0.15 inch is wasted as far as desert plants are concerned, for the moisture evaporates before penetrating the soil. Some annuals produce seeds having water-soluble germination inhibitors in their coverings, hence fail to sprout, even after rain, unless the moisture totals at least half an inch.

When soil moisture from December and January rainfall is enough to support potential plants it dissolves the seed coats, and the desert floor is soon carpeted with eager green seedlings. When winter rains are scant, as is so often the case, the dormant seed population fails to germinate and the spring flower display doesn't appear. There is no sure way to forecast a spectacular blossom year, for a sudden cold wave or period of drying winds may literally nip in the bud a potential season of brilliant bloom. A great flower year may occur only once in a decade.

Perennials are more dependable than annuals, since some of them, particularly cactuses and other succulents, have water storage tissues in their stems or roots. These perennials may be counted on to blossom each year, but with much less abandon than after winters of above normal precipitation. Many perennials have surprisingly extensive root systems.

Fascinating are the ways by which plants have managed to thrive under severe conditions of desert heat and drought. As we have seen, most annuals lie dormant as seeds until suitable moisture and temperature occur. Then they grow very rapidly, to bloom and mature seeds while the soil still has moisture. Winter rains produce the spring-blooming ephemerals, and summer showers produce the summer "quickies."

Another group of plants, including the ocotillo (oh-koh-TEE-yoh), slows down the life processes and becomes dormant during dry periods, even to dropping all their leaves. When rains come they put on new leaves, several times a year if necessary.

Cactuses and other succulents gorge themselves with water when the soil is wet, releasing the moisture very sparingly from storage tissues during the "long dry." Some have discarded or reduced their foliage, or have covered leaf surfaces with varnish or wax, to decrease to a minimum the loss of vital moisture through transpiration.

Identifying Desert Wildflowers

Unless you are a botanist, identification of flowers by measuring and counting their various parts, as described in technical keys, is generally too complicated to be practical. Several years ago, recognizing this problem, I authored a book, *Flowers of the Southwest Deserts,* illustrated by Jeanne R. Janish and published by the Southwestern Monuments Association, designed to aid the wildflower fancier in plant identification by color-grouping the flowers. With Mrs. Janish's superb illustrations pointing out each plant's most obvious characteristics, it has proved an excellent field guide. However, the demand for natural color flower portraits could not go unheeded, and this book is the result. The two books complement each other, although each fills a need in its own right. Used together, they make you more positive of some identifications.

Probably the best way to become acquainted with a flower is to be introduced to it by someone. But there is one catch to this method — one plant may be known by many aliases.

When the Spaniards came into the Southwest over 400 years ago they found the Indians had names for some of the flowers in their own languages. The Spaniards added their own names, and later the Americans added English names. Some of these names were those of similar-appearing but quite different flowers they had known "back East." Later, scientists came to study the desert plants, and gave them all Latin names.

To assist in standardizing names of desert flowers, this booklet gives preference in its headings to common and scientific names found in *Standardized Plant Names,* Second Edition, 1942. Tree names, both common and scientific, follow the *Checklist of Native and Naturalized Trees of the United States,* by Elbert L. Little, Jr., 1953. For plants whose names do not appear in either of the foregoing, we have followed *Arizona Flora,* by Kearney and Peebles, Second Edition, 1960. In addition, placed within the text, are listed some of the better-known common names that we have met.

There are hundreds of desert flowers, some quite common, for which there was not space in this booklet. If you wish to broaden your acquaintance to include more of them, we recommend for added reading the publications listed in the back of this book.

The author wishes to express here sincere thanks to Mrs. Polly Patraw for invaluable assistance in identifying a number of the flowers whose pictures appear in this booklet.

Spring gives an Evening Party

When Paloverde trims her golden gown,
And Deerhorn dons her filaments of white;
When tall Saguaro fits his fragrant crown
In preparation for the party night;
When bats across the ruby sunset dance,
When Ocotillo lights his candle's flame,
When verdure carpets Desert's wide expanse,
Then Spring is in the Southwest once again.

The linnets in their scarlet vests and caps
Are first to answer Spring's insistent call,
While white-crowned sparrows scan their travel maps,
Discussing details of the coming ball.
Then thrashers practice every morn and eve
The songs they'll sing upon that night of nights,
While phainopeplas, in their haste to leave,
Dash back and forth in short, impatient flights.

The desert halls glow bright as time draws near.
Each cactus wears her frilled and perfumed dress.
Ground squirrels, for this joyous time of year,
Sport their best furs. The rabbits do no less.
From far and near the desert folk have come
To 'wait their hostess, Spring, who, very soon,
Will lift stars o'er the skyline, one by one,
And then turn on the glorious, golden moon.

1. Longleaf ephedra

Commonly called "Mormon tea," there are many species of ephedra (ef-FED-rah) growing throughout the Southwest. This yellow-green, stringy-stemmed shrub with tiny, scale-like leaves, is usually 3 to 4 feet tall, but sometimes reaches a height of 12 feet. Its small, fragrant, springtime flowers grow in dense clusters that attract insects. Some species provide winter forage for cattle and are said to be browsed by bighorn sheep. Pioneers brewed a palatable drink from the dried stems. Certain Indian tribes considered the brew a tonic, beneficial for treatment of syphilis and other diseases. The drug, ephedrine, comes from a Chinese member of this genus.

Ephedra trifurca   Jointfir Family

LONGLEAF EPHEDRA

COMMON REED

2. Common reed

Somewhat resembling bamboo, *carrizo* grows in dense thickets in marshes, along river banks, and in other wet locations. Largest of the grasses, it sometimes attains a height of 12 feet. The large, tassel-like flower heads appear from July to October and create a spectacular mass display. The horizontal rootstalks interlock, crowding out other plants. A single rootstalk may extend 30 feet. The straight, hollow stems served Indians as arrowshafts, pipestems, and loom rods. Along the Mexican border the leaves are woven into mats and the long, sturdy stems are used as screens and in roofing native houses.

Phragmites communis   Grass Family

3. Spiderwort

Because of its slender, drooping leaves, this delicate blue-to-violet, three-petaled flower might easily be mistaken for a lily. Plants grow from 8 to 18 inches high. A perennial, the spiderwort's thick, succulent roots enable it to produce blossoms from April to September. Not abundant, it is usually found in moist locations in desert mountain ranges at elevations above 2,500 feet. Flowers form in clusters at the tip of a plant's stem, and are pollenized by bumblebees that eat the pollen.

Tradescantia occidentalis   Spiderwort Family

SPIDERWORT

DESERTLILY

4. Desertlily

Limited in its range to the desertlands of southern California and southwestern Arizona, the desertlily or ajo (AH-hoe) resembles a small easter lily. During dry seasons the plants do not bloom, but following wet winters each deeply-buried bulb sends up a vigorous shoot which may be from 6 inches to 2 feet tall, with a bud cluster at its tip. The delicately fragrant flowers may appear in late February, with some tardy bloomers still in evidence in early May. Bulbs were dug and eaten by Indians and, because of their flavor, were called *ajo* (garlic) by the Spanish pioneers. The town of Ajo and a nearby valley and mountain range in southwestern Arizona were named for this plant.

Hesperocallis undulata   Lily Family

5. Weakstem mariposa

Similar in appearance to the segolily, State flower of Utah, weakstem mariposa, sometimes called "straggling butterfly lily," varies in color from white to pale purple. The slender stem is not erect, like other mariposas of which there are many species, but wanders over the ground or makes its twisting way among the branches of low shrubs. It grows at elevations up to 4,000 feet on slopes and benches of mountains of the Mojave-Colorado Desert, in the Death Valley area, and in the desert mountains of southern Arizona, blossoming during April and May. Indians and pioneers ate the bulbs.

Calochortus flexuosus   Lily Family

6. Golden segolily mariposa

Considered by some botanists as a distinct species, this mariposa or "butterfly tulip" is found in the higher mountains of the eastern Mojave-Colorado Desert and also in the vicinity of the Painted Desert of northern Arizona. Common in Petrified Forest National Park from May to July, the bright yellow flowers make an eye-catching display among the colorful pieces of petrified wood covering the ground. The bulbs can withstand severe cold, but suffer during winters when there is frequent freezing and thawing.

Calochortus nuttalii aureus   Lily Family

DESERT MARIPOSA

7. Desert mariposa

Brightest of the mariposas, the short-stemmed, flame-like flowers usually appear singly, but may occur in patches, producing in April a spectacular display visible from a long distance. Plants growing under bushes elongate their stems to elevate their blossoms into the sunlight. Occasional in the Mojave-Colorado Desert, this species is abundant in the foothills of some of southern Arizona's mountain ranges, exceeding even the goldpoppy in the neon-like brilliance of display. *Mariposa* is Spanish for butterfly, and the genus name *calochortus* is Greek for beautiful grass.

Calochortus kennedyi   Lily Family

8. Soaptree yucca

Common throughout the Southwest, the many species of yuccas (YUH-kuhs) are of two major groups, the narrow-leaf and the wide-leaf. Called "soaptree" because of its height (maximum 30 feet) and the fact that its roots contain saponin, soaptree yucca or *palmilla* (pahm-EE-yah — "little palm") belongs in the narrow-leaf group. From southwestern Arizona across southern New Mexico, and from west Texas southward into the Mexican states of Chihuahua and Sonora, this spectacular plant blossoms in May and June on desert grasslands from 2,000 to 6,000 foot elevations. Cattle eat the young flower stalks, and Indians used the leaf fibers for making fabrics, basketry, and other items. The yucca is the State flower of New Mexico.

Yucca elata   Lily Family

SOAPTREE YUCCA

9. Joshuatree yucca

Another of the narrow-leaf yuccas and largest of the genus, the joshuatree is restricted in its range to the Mojave-Colorado Desert, of which it is the principal indicator. Blossoms, which do not open as wide as those of other species, grow in tight clusters at the tips of the branches, appearing in March and April. Joshuas do not blossom every year, the interval between flowerings depending upon rainfall and temperature. A small night lizard is dependent upon the joshuatree, at least 25 species of birds nest among its branches, and a prehistoric mammal, the yucca-feeding ground sloth *Nothrotherium,* lived within the plant's then wider range. Joshua Tree National Monument, in south-central California, was established in 1936 to preserve spectacular forests of this tree yucca.

Yucca brevifolia   Lily Family

10. Torrey yucca

Unlike the narrow-leaf soaptrees which produce dry, capsular fruits, the wide-leaf yuccas bear fleshy fruits which Indians cooked and ate. Indians also used the leaf fibers in weaving fabrics. Roots contain saponin and the Indians still cut them up and use the pieces for soap, especially as a shampoo. The stiff, fleshy leaves with needle-sharp tips give the plant the name "Spanish bayonet." Torrey yucca blooms in April in southeastern New Mexico and west Texas, with similar plants, *Yucca schottii* in southern Arizona, and *Yucca schidigera* in the Mojave-Colorado Desert.

Yucca torreyi   Lily Family

11. Carneros yucca

Massive and thick-stemmed, the locally-named "giant dagger" is supposedly limited in its native range in the United States to Brewster County, Texas. A colony resembling this species has been reported recently in McKittrick Canyon in the Guadalupe Mountains. An extensive forest of these spectacular plants has given the name Dagger Flat to a broad valley in the Sierra del Carmen of Big Bend National Park. Usually blossoming in April, the massive, white flower clusters gracing the crowns of thousands of these majestic yuccas create a never-to-be-forgotten spectacle. A small night-flying moth is the yuccas' pollenizing agent and, in return for this essential service, lays her eggs in the plants' ovaries where the young feed on the developing seeds.

Yucca carnerosana   Lily Family

CARNEROS YUCCA

12. Sacahuista

Sometimes confused with the yuccas, the several species of "beargrass" or "basketgrass" have pliant, grasslike leaves, small flowers, and papery fruits. The plumelike blossom panicles open in May and June. The plants favor rocky hillsides, and rarely occur on valley floors. Indians roasted the tender bud stalks for food, and cattle browse the leaves when other vegetation is lacking. Mexicans, in weaving basketry, use the entire leaves, which are especially desirable for fashioning basket handles.

Nolina microcarpa   Lily Family

SACAHUISTA

WHEELER SOTOL

AGAVE

13. Wheeler sotol

Also likely to be confused with the yuccas, sotol has a basal cluster of pliant, ribbonlike leaves edged with hooked thorns, and a tall flower stalk bearing at its upper end a dense panicle of small, creamy (sometimes brown) flowers. Blossoming in May and June, the maturing flower clusters remain attractive throughout the summer. Mexicans split the succulent basal crowns and allow the sap to ferment, producing the fiery alcoholic beverage, sotol (SOH-tole). Desert-dwelling bighorn sheep are said to browse the tough leaves. The stiff leaf bases, when pulled from the cluster, form the "desert spoons" sold in some curio stores.

Dasylirion wheeleri   Lily Family

14. Agave

Many species of agaves (ah-GAH-vees) or "century plants" attract attention on desert hillsides when they send up their tall blossom stalks in June and July. The thick, fleshy, sharp-tipped leaves form a basal rosette. Some of the larger species may require 10 to 20 years to store enough plant food to produce the sturdy, fast-growing flower stalk. After blossoming, the exhausted plant dies. *Agave scabra,* one of the spectacular forms, is limited in its range to the Chisos Mountains of Big Bend National Park, Texas.

Agave scabra   Amaryllis Family

15. Parry agave

Another of the large "century plants," Parry agave blooms from June to August, producing spectacular displays on hillsides in northern Mexico, southern New Mexico, and southern Arizona. Some of the larger agaves are called mescal (mess-KAHL) because of a potent alcoholic beverage of that name distilled from the fermented sap derived from the bud stalks. Tequila (tee-KEEL-ah), the famous native drink of Mexico, also is distilled from fermented agave juices, and the beerlike pulque (pool-KAY) has a similar derivation. Indians roasted the bud stalks in stone-lined pits covered with hot rocks. Some of these pits may still be seen.

Agave parryi   Amaryllis Family

16. Lechuguilla agave

One of the common plants of the Chihauhuan Desert and considered the principal indicator of that region, lechuguilla (lay-chu-GHE-ah) covers the ground so densely in some places that it is impossible to walk through it. The stiff, erect, needle-tipped, banana-shaped leaves are a hazard to man and beast. The flowering stalk, which blossoms in May and June, is unbranched and flexible, bending gracefully in the desert breeze. Deer and cattle nip off the tender buds. Mexicans weave the tough leaf fibers into coarse fabrics; and the roots, called *amole,* produce suds when rubbed in water.

Agave lophantha poselgeri   Amaryllis Family

PARRY AGAVE

LECHUGUILLA AGAVE

17. Canaigre

This coarse, herbacious perennial is one of the early spring flowers of the desert, sometimes blooming along road shoulders and in sandy washes in late February and March. Commonly called wild rhubarb, its sap and roots are high in tannin content, and its delicately pink fruits are more attractive than the blossoms. Indians and Mexicans use the leaves for greens. Papago Indians of Arizona roast the leaves and use the roots for treating colds and sore throat. This plant is a close relative of European dock, several species of which have become naturalized in North America.

Rumex hymenosepalus   Buckwheat Family

18. Trailing allionia

Blossoming from April to October, trailing allionia, known in some places as "trailing four o'clock" or "windmills," is a spreading annual with small but colorful blossoms on long, trailing stems. The prostrate branches are sticky, so are often covered with grains of sand and flecks of mica. What appears to be one blossom is actually three flowers, giving it the name "pink three-flower." It is found on dry, sandy benches throughout desert regions of the Southwest. Fruits are winged.

Allionia incarnata   Four o'clock Family

19. Desert sandverbena

One of the early spring flowers, sandverbena creates spectacular mass displays, sometimes alone, usually intermingling with other colorful early bloomers such as bladderpod and sundrops, which grow on road shoulders and sandy flats. The flowers are delicately fragrant, especially at night. Semi-prostrate in habit, sandverbena leaves are covered with a dense growth of short, soft hairs which retard the loss of moisture so essential to desert plants. This annual is common from southern California and southern Arizona into Sonora.

Abronia villosa   Four o'clock Family

20. Mexican goldpoppy

Closely related to the orange California-poppy, official flower of the Golden State, the desert species is a bright yellow annual. Following warm, wet winters clusters of these glorious blooms dot the hillsides in late February or early March. By April they may cover the slopes with a blanket of gold interwoven with the blue threads of lupines and purple patches of escobita owlclover. When other early spring vegetation is scarce, cattle graze the plants. Flowers open only during sunny hours, remaining tightly closed at night and on cloudy days.

Eschscholtzia mexicana   Poppy Family

21. Crested pricklepoppy

Not restricted to a desert habitat, this spiny-leafed perennial is widespread on dry soils from Nebraska to Wyoming and from Texas to southern California and Mexico. Abundant throughout the summer, the flowers may be found, in warm climates, during every month of the year. Copious spines and the acrid yellow sap make the plants distasteful to cattle, so a heavy growth of pricklepoppy may be an indicator of an overgrazed range. Also called "thistiepoppy," a single plant may be graced by a dozen or more fragile flowers, each ready to be replaced by one or more prickly buds. The seeds are said to contain a powerful narcotic.

Argemone platyceras   Poppy Family

22. Desert bearpoppy

Limited in its range to the Mojave-Colorado Desert, and with much smaller flowers than either the goldpoppy or pricklepoppy, the desert bearpoppy is, nevertheless, a handsome plant when in bloom. Usually flowering in April with creamy to light yellow blossoms, the foot-high plants often cover large expanses of open desert. This mass of pale color may be broken by splashes of bright red where patches of beavertail pricklypear mark small rocky islands, or where ocotillos wave their scarlet-tipped wands in the spring breeze.

Arctomecon meriamii   Poppy Family

23. Spectaclepod

Found at elevations above 1,000 feet, spectaclepod is one of the long-flowering species blooming from February to October. The large flower heads are pleasantly fragrant, and the peculiar, flat, double fruits resemble tiny spectacles protruding at right angles to the stem. This species is found in the Petrified Forest area of northern Arizona, and Hopi Indians are reported to use the plant in treating wounds. Another species, California spectaclepod, is often abundant, covering sandy flats of the lower deserts. This species blooms from February through April and sometimes again in the fall.

Dithyraea wislizenii   Mustard Family

24. Bladderpod

Another early bloomer, February to May, bladderpod is one of the first spring flowers to spread its yellow carpet across the desert flats. The small, low-growing plants lift numerous clusters of four-petaled flowers, forming an understory of color among the taller herbs. In some localities, bladderpods are called "beadpods" because of the spherical fruits. The plants afford good forage for cattle. A close relative, with white to purple flowers, is found from Texas to Arizona and Mexico, starting to blossom in January during warm winters.

Lesquerella gordonii   Mustard Family

25. Coast erysimum

A showy plant with a large terminal cluster of four-petaled flowers, it is frequently called "western" or "desert wallflower." When growing under shrubs it often extends its stems 2 feet or more to reach up into the sunshine. Usually blossoming in March, some plants may be found blooming at almost any time during the summer to as late as September.

Erysimum capitatum   Mustard Family

COAST ERYSIMUM

FALSE-MESQUITE CALLIANDRA

26. False-mesquite calliandra

With mimosa-like leaves and long-stamened flowers growing in clusters, false-mesquite calliandra or "fairy duster" is a small, straggling bush, quite Japanesy in appearance, from a few inches to 3 feet high. It blossoms from February to May, and is quite common below 5,000 feet from west Texas to southern California and northern Mexico. In California it is especially abundant along the east side of the Chocolate Mountains. During periods of drought the leaves enter a state of continued wilt, but revive promptly when rain comes.

Calliandra eriophylla   Pea Family

CATCLAW ACACIA

27. Catclaw acacia

Also known by such descriptive names as "tear-blanket" and "wait-a-minute," catclaw acacia is one of the notoriously thorny shrubs or small slender trees of the rocky hillsides and borders of desert washes. Flowers are fragrant and, during the blooming period in May, attract many insects, including honey bees, which gather and store nectar that makes high quality honey. The stringbean-like fruits turn red in late summer and, if abundant, make a spectacular show. These fruits were ground into meal and used for food by Arizona and Mexican Indians. Thickets of catclaw acacia provide havens of refuge for birds and rabbits pursued by hawks or other predators.

Acacia greggii   Pea Family

28. Mescat acacia

Armed with long, slender, straight white spines, giving it the name "white-thorn," this pretty flowering shrub is abundant over large areas of dry slopes and mesas from Texas to Arizona and Mexico at 2,500 to 5,000 feet. It is often used as a decorative in landscape plantings around buildings. Blossoms are fragrant and sometimes continue from May to August; the shrub occasionally blooming again in November. Cattle and horses eat the bean-like fruits.

Acacia constricta   Pea Family

MESCAT ACACIA

29. Honey mesquite

Mesquite (mess-KEET) is a many-branched tree 15 to 25 feet tall, which flowers from late April to June. It is common bordering desert washes, often forming dense thickets. The flowers furnish honey bees and other insects with nectar, and the long, sweet pods ripen in autumn, providing food for livestock. The fruits have long been a staple in the diet of desert Indians, who used the trunks, roots, and branches of the trees for firewood and the dried gum-like sap to mend pottery and as a black dye. The inner bark provided the Indians with materials for basketry and coarse fabrics. Roots of mesquite trees have been reported to penetrate to a depth of 50 to 60 feet to tap sources of ground water.

Prosopis juliflora   Pea Family

30. Senna

Blossoming on rocky slopes and mesas from April to August, this species is quite common from Texas to Arizona and Mexico. A close relatives, *Cassia armata,* represents the genus in western Arizona, southern Nevada and southeastern California, where it blooms in a riot of color in April and May following wet winters. Senna is sometimes called "rattlebox," because the nearly ripe seeds rattle in their woody pods when the plant is stirred, startling the hiker who thinks "rattlesnake!" The plants often grow in large clumps 2 feet high and 2 or 3 feet across, with little foliage.

Cassia bauhinioides   Pea Family

31. Blue paloverde

Perhaps the most dependable of spring bloomers, blue paloverde trees cover themselves with masses of yellow blossoms in April and May. Usually found alongside desert washes, they mark these ephemeral stream courses as paths of gold threading the open desert. During much of the year the trees are relatively leafless, the green bark of trunk and branches taking over the function of leaves. The word *paloverde* (PAH-low-VEHR-dee) means "green stick" in Spanish, refering to the color of the bark.

Cercidium floridum   Pea Family

32. Paradise poinciana

Not a southwestern desert native, this striking shrub, 3 to 10 feet high, usually called "bird-of-paradise-flower," was introduced from South America and has escaped from cultivation to establish itself in parts of the desert where conditions are suitable. The blossoms are showy but ill-smelling, and are popular as ornamentals about homes, especially in Mexico. The shrub's principal advantage in landscape plantings is its long blossoming period, which sometimes lasts from April to September.

Poinciana gilliesii   Pea Family

33. Coulter lupine

This is but one of many species of lupine, both annual and perennial, common throughout the West at nearly all elevations. Perhaps the most publicized is the "Texas" lupine, or "bluebonnet," hailed by Texans as their State flower. Desert species are early bloomers, sometimes appearing in protected sandy soils and on highway shoulders in January. In favorable seasons masses of these handsome blue to violet blossoms color desert hillsides with acres of fragrant bloom. Sometimes growing in pure stands, often mixed with a variety of other spring flowers, lupines may usually be found blossoming as late as June.

Lupinus sparsiflorus   Pea Family

COULTER LUPINE

INYO LUPINE

34. Inyo lupine

Considered one of the more handsome of the desert perennials, the "adonis" lupine, as it is known in southern California, is found near sandy washes in the high desert. It is especially abundant in Joshua Tree National Monument. The name *adonis* refers to its great beauty. The name *lupinus* is derived from the Latin *lupus* meaning wolf, because these plants were at one time thought to be soil predators. Actually, as with other members of the pea family, lupines are able to take atmospheric nitrogen and leave it in the ground, thereby increasing rather than depleting soil fertility.

Lupinus excubitus   Pea Family

SMOKETHORN

35. Smokethorn

Better known as "smoketree," this silvery-gray, seemingly leafless shrub grows in and along sandy washes below 1,500 feet, throughout the Mojave-Colorado Desert. At a distance it resembles a plume of smoke rising from a campfire. Its small but violet to indigo flowers cover it with a gorgeous blue blanket in May, making it one of the really handsome desert shrubs. It requires ample supplies of water, hence is restricted to washes that carry runoff from both winter rains and summer downpours. The seeds sprout readily, and the seedlings with their well-formed leaves look very unlike their parents. Few seedlings survive the hazards of drought or being smothered by sand carried down the washes by flash floods following cloudbursts.

Dalea spinosa   Pea Family

36. Fremont dalea

Noted for its royal purple flowers, this low shrub, less than 3 feet high with peculiar zig-zag branches, blossoms from April to June. In common with other daleas (day-LEE-ahs) it is usually called "indigobush" or "peabush." It is normally found below 3,000 feet in desert mountain ranges from southern Utah through Arizona and southeastern California. There are many species of dalea in the desert, all characterized by deep blue to indigo and rose-violet flowers, which attract attention by their beauty. Indians used the extract from twigs for dyeing basketry.

Delea fremontii   Pea Family

FREMONT DALEA

TESOTA

WOOLLY LOCO

37. Tesota

Thriving only in a frost-free climate, this is among the largest and most beautiful of desert evergreen trees. It is usually found along sandy washes, mingling with mesquites and paloverdes. It is particularly susceptible to mistletoe infestation, which has killed or weakened many fine trees. Blossoming in May and June, the trees are sometimes laden with lavender, wisteria-like flowers. The wood is extremely hard and heavy, hence the tree is locally known as "ironwood," or *palo-de-hierro,* in Mexico. Indians ate the seeds and used the wood for tool handles and arrow-points. Its long-burning qualities made it especially desirable for fuel. As a result, many of the trees have been cut, making it one of the species threatened with extinction.

Olneya tesota Pea Family

38. Woolly loco

Many species of "locoweed" ranging in color from deep purple to creamy white are found throughout the desert at nearly all elevations. They sometimes create extensive mass displays but are more commonly found mixed with other flowers. Species with bladder-like pods are called "rattleweed." Loco in Spanish means "crazy" and refers to the fact that a number of species of *astragalus* contain selenium, which causes a serious disease among livestock, especially horses, that eat it and as a result "act crazy."

Astragalus mollissimus Pea Family

39. Texas heronbill

Both this species and its close relative, al-fileria (*Erodium cicutarium*) are early blossoming annuals often widespread on plains and mesas, February to May. The flowers, although abundant, are small and so hidden in the low-growing foliage that they rarely create a mass display. "Filaree," as these plants are locally known, is not native, but is believed to have been introduced from Europe by the Spaniards, and is now naturalized throughout the Southwest. The corkscrew-like appendages of the fruits are tightly twisted when dry, but untwist when moist, literally screwing the sharp-pointed fruits into the soil. Both species provide excellent spring forage for livestock.

Erodium texanum   Geranium Family

40. Coville creosotebush

Often erroneously called "greasewood," creosotebush is generally recognized as the most adaptable of all desert plants, and a definite indicator of the Lower Sonoran Life Zone. The shrubs cover thousands of square miles, often in pure stands, and flower throughout much of the year, but most profusely in April and May. Fuzzy white, globular fruits are almost as spectacular as the flowers. The plant can endure long periods of drought. Following rains its foliage gives off a musty, resinous odor, suggestive of creosote, stimulating the Mexican name *hediondilla* (little stinker). In Mexico the plant is considered to have medicinal values and many uses. The Pima Indians boiled the leaves, using the decoction as an emetic and to poultice sores. They used the lac, found as an incrustation on the branches, to cement arrowpoints and to mend pottery.

Larrea tridentata   Caltrop Family

TEXAS HERONBILL

COVILLE CREOSOTEBUSH

CALTROP

41. Caltrop

Often abundant on road shoulders and in low spots where rainwater from hot-weather showers provides adequate moisture, caltrop or "summerpoppy," with large blossoms and attractive compound leaves, decorates the desert when other flowers are noticeable by their absence. The long, weak stems, usually prostrate, give the plants a vine-like appearance, but when growing under shrubs they extend upward so that the shrub is mistakenly thought to be blooming. Superficially resembling the springtime goldpoppy, caltrop has five rather than four petals, and may be found in bloom as late as October.

Kallstroemia grandiflora   Caltrop Family

DESERT GLOBEMALLOW

42. Desert globemallow

Ranging in size from delicate 6-inch annuals to coarse, woody perennials 4 feet high, the globemallows vary in color from creamy white to pink, rose, peach, and lavender. Desert globemallows flaunt their graceful, blossom-covered stems along roadsides or on the banks of sandy washes. Because some people are allergic to them, globemallows are called "sore-eye poppies" in parts of southern Arizona, and in Lower California are known as *plantas muy malas* (very bad plants).

Sphaeralcea ambigua   Mallow Family

43. Five-stamen tamarisk

Sometimes confused with tamarack because of the similarity of names, five-stamen tamarisk, locally called "salt-cedar," is one of several small tree species from southeastern Europe and western Asia which have become naturalized in North America. "Salt-cedar" often forms dense thickets on alkaline soils along stream and reservoir banks at elevations below 5,000 feet. Flowers, which vary in hue from deep pink to white, cover the trees with graceful plumes of color from March to August. Although valuable in retarding soil erosion, tamarisk requires large quantities of water, an especially undesirable characteristic in the arid Southwest.

Tamarix pentandra   Tamarix Family

44. Samija mentzelia

There are many kinds of mentzelia, all herbs found in the West. The barbed hairs which cover the stems and leaves cause the plant to cling to whatever it touches, giving it the common name "stick-leaf." Flowers grow at the ends of the branches and some species open fully only in sunlight. *Mentzelia involucrata,* also called "sand blazing star," is an annual, 4 to 16 inches high, found in hot sandy washes below 3,000 feet in southwestern Arizona, southeastern California, and northern Sonora. It blooms from February through April. One variety, *megalantha,* has larger brighter, yellower flowers than the typical *involucrata.*

Mentzelia involucrata   Loasa Family

45. Stingbush

Because it is usually found growing from crevices in cliffs, this low, rounded bush is often called "rock-nettle." When covered with large blossoms from April to September the plant has a striking appearance. The pale green leaves are covered with stinging hairs, strong enough to impale such small creatures as bats emerging from cave entrances where these plants are growing. Stingbush is common in the desert ranges of southeastern California, especially in the Death Valley area, to western Arizona and southern Nevada.

Eucnide urens   Loasa Family

46. Deerhorn cactus

Easily overlooked, when not in blossom, as a group of slender, fluted, gray-green stems hidden beneath a shrub, the beauty and fragrance of the deerhorn cactus flowers have earned it the name, in Mexico, of *reina-de-la-noche,* meaning "queen-of-the-night," and in the Southwest it is commonly called "night-blooming cereus." Buds unfold soon after sunset in late June or early July, perfuming the desert air and attracting night-flying insects. They wilt soon after sunrise the following morning. The large, tuberous root, which serves as a water-storage organ, usually weighs from 5 to 15 pounds, but specimens have been found weighing more than 80 pounds. Indians at one time dug the tubers for food. The bulbous fruits become red when mature, and are almost as spectacular as the flowers. This species is found from west Texas to western Arizona and northern Mexico.

Peniocereus greggii   Cactus Family

SAGUARO

47. Saguaro

Largest of the cactuses in the United States, the saguaro (suh-WAR-oh) is limited in its principal range to southern Arizona and northern Mexico. Although rarely exceeding 30 feet in height, specimens 50 feet tall and weighing up to 10 tons, are on record. Blossoms form as huge bud clusters at the branch tips, opening a few at a time each night, usually in May, and remain open until mid-afternoon of the following day. Fruits of the saguaro are eaten by birds and other animals, and at one time were important in the diet of desert Indians. The state flower of Arizona and the subject of a U.S. postage stamp issued in February 1962 to commemorate the 50th anniversary of Arizona's statehood, the saguaro is also commemorated and protected in the National Monument of that name near Tucson.

Cereus giganteus   Cactus Family

48. Organpipe cactus

ORGANPIPE CACTUS

Limited in its range to northwestern Mexico and the vicinity of Organ Pipe Cactus National Monument in southwestern Arizona, this columnar cactus grows in clumps of spine-covered stems, some of which may be 10 to 15 feet in height, rarely branching, and with no central trunk. Blossoms open at or near the stem ends during May nights, and close the following day. The spine-covered fruits, about the size and shape of a hen's egg, have long been harvested by the Papago Indians, who boil the sweet juice to the consistency of syrup and store the pulp and seeds for winter food. The fruits are locally called *pitahaya dulce,* or sweet cactus fruit.

Lemaireocereus thurberi   Cactus Family

49. Claretcup echinocereus

Not only are there many species of *Echinocereus,* popularly called the "hedge-hog cactuses," but there are also several varieties of *Echinocereus triglochidiatus.* One variety sometimes develops into cushion-like mounds composed of several hundred oblong stems huddled together with a seemingly precarious foothold in crevices among the rocks or on rocky slopes of the Mojave desert. Another grows in loose clusters of cylindrical stems in the higher desert grasslands up to the oak belt in the mountains of southern New Mexico, Arizona, and northern Mexico. When blossoming in May and June these clustering "hedgehogs" create a spectacular display.

Echinocereus triglochidiatus   Cactus Family

50. Engelmann echinocereus

One of the more common species of "hedghog," sometimes called "strawberry cactus," the Engelmann echinocereus grows as 2 to 12 or more robust, cylindrical stems up to a foot in height, among the creosote bushes and bur-sages of the Sonoran and Mojave-Colorado Deserts, flowering from February to May. Flowers close at night and reopen the following morning. Blossoms vary considerably in color from purple to lavender. Spines, too, are variable, from gray and yellow to dark brown. In southeastern California, where it is common, this species is called "calico cactus" because of its many-colored spines. Fruits of some varieties (of which there are many) are edible, forming an important item in the diet of birds and rodents.

Echinocereus engelmannii   Cactus Family

RAINBOW ECHINOCEREUS

51. Rainbow echinocereus

Far from common but among the more beautiful of the "hedgehogs" is the rainbow echinocereus, also called "rainbow cactus," so named because of the horizontal bands of alternating red and white spines encircling the single, sturdy stem. It grows in rocky situations in the mountains of southern Arizona and northern Mexico, blossoming from June to August. The large flowers, of which there may be from one to four crowding around the crown of the plant, are often larger than the plant itself. Spines are small and lie densely flat over the somewhat fluted stem, which is from 3 to 14 inches high.

Echinocereus rigidissimus   Cactus Family

YELLOW PITAYA ECHINOCEREUS

52. Yellow pitaya echinocereus

Sometimes called "Texas golden rainbow," the yellow pitaya of the Chihuahuan Desert is similar in appearance, except for the color of its blossoms, to the rainbow echinocereus. Quite common in portions of Big Bend National Park, the stubby, upright stems usually grow singly but sometimes occur in small clusters. The term *pitaya* or *pitahaya* is commonly applied along the Mexican border to cactuses bearing edible fruits. In Texas the term refers to the low-growing floral hedgehogs; in Arizona to the columnar cactuses. Pricklypear cactuses having soft, juicy, edible fruit are known as *tunas*.

Echinocereus dasyacanthus   Cactus Family

53. Southwest barrel-cactus

Massive, cylindrical, and covered with clusters of stout spines, the central one hook-shaped, these desert giants are often mistaken for young saguaros. There are several species, all locally called *bisnagas,* with some quite small and others attaining a height of 5 or 6 feet. The majority produce clusters of orange to red flowers on their crowns in late summer, but the yellow-flowered California barrel-cactus blossoms in the spring. Their tendency to lean toward the light causes many of these heavy-bodied plants to tip in a southwesterly direction, giving them the name "compass cactus." This group is naively believed by some people to contain water. Actually the slimy, alkaline sap obtained by mashing the pulpy flesh might conceivably save someone lost in the desert from dying of thirst. The pale yellow fruits are not spiney, and are eaten by birds, rodents, deer, and other desert animals.

Ferocactus wislizenii   Cactus Family

54. Fishhook mammillaria

There are a number of species of the low-growing, usually dome-shaped mammillarias, the solitary kinds often so small as to be overlooked except when blooming, in late spring or early summer. Some are known as "fishhook cactuses" because of their long, slender, hooked spines, others as "pin-cushion cactuses" because of the shape of the plants. The large, colorful blossoms which encircle the stems mature usually to red, in some species green, nipple-shaped fruits. Members of this genus are widespread in grasslands or rocky mesas and slopes throughout the Southwest.

Mammillaria microcarpa   Cactus Family

55. Beavertail pricklypear

Limited in its principal range to the Mojave-Colorado Desert, the beavertail is a low-growing species with flat joint-pads and bluish-green stems without spines. In their place are clusters of brownish spicules set in slight depressions in the wrinkled pads. The plants blossom in March and April, adding materially to the color of the spring flower display. The plants thrive in sandy desert soils, at elevations from 200 to 3,000 feet above sea level, and are found as far east in Arizona as Wickenburg. Cahuilla Indians cook the fruits with meat, and Panamint Indians dry the pads and boil them with salt.

Opuntia basilaris   Cactus Family

56. Engelmann pricklypear

Most widely distributed of the pricklypears, Engelmann plants are large and spreading, sometimes forming spiney bushes 3 to 5 feet high and up to 15 feet in diameter. The branching stems may have from 5 to 12 pad-joints. Flowering in April and May, the petals at first are yellow but turn to pink or rose with age. The plants prefer washes and benches in the desert grasslands, often growing with paloverdes, saguaros, mesquites, and lechuguilla agaves. Excessive abundance often indicates an overgrazed range. Fruits, called *tunas,* are purple to mahogany when mature, and are eaten by many birds and rodents, as well as by desert Indians.

Opuntia engelmannii   Cactus Family

57. Arizona jumping cholla

Also known as "silver cholla" (CHOY-uh) and "teddybear cactus," this stocky bush cactus, with a short sturdy trunk and compact, densely spined crown, is common on hot rocky, south-facing hillsides. Joints are extremely brittle and the barbed spines catch so easily in the hair of animals or clothing of persons that the joints appear to jump from the plant. Joints broken off by the wind fall to the ground and take root in the sandy soil, gradually developing forests of this striking cactus, easily recognized by the silvery sheen of the spines. The attractive flowers which appear from March to May blend inconspicuously with the spiney joints.

Opuntia bigelovii   Cactus Family

58. Tesajo

Common along banks of washes and on desert flats, the tesajo, or "Christmas cholla," is so slender-stemmed and sprawling in growth habit that it is easily overlooked in a tangle of vegetation. Its flowers, appearing in May and June, are small and inconspicuous, but the orange to scarlet fruits about the size and shape of olives, are striking eye-catchers in the fall and winter months. In the open the shrubby plants are rarely more than 2 feet high, but in thickets of northern Mexico some have become almost vinelike, growing up through mesquite or paloverde trees to a height of 12 feet or more. The species grows at elevations of 200 to 5,000 feet from Texas to western Arizona and northern Mexico.

Opuntia leptocaulis   Cactus Family

59. Whipple cholla

This low-growing cholla of the higher desert above 3,500 feet, is characteristic of the plateau grasslands, forming mats of short but erect stems usually less than 2 feet high. It blossoms in June and July. The tender young stems and yellow, fleshy fruits are browsed by pronghorns, and the fruits are also used by the Hopi Indians for food and as a seasoning. Because of its customary low-growing habit it is something of a hazard to hikers. It is considered the most widely distributed cholla in Arizona.

Opuntia whipplei   Cactus Family

60. Walkingstick cholla

Flowering in May and June and common throughout southwestern New Mexico, southern Arizona, and northern Mexico, the walkingstick cholla is best known because of its persistent clusters of yellow fruits. These remain throughout the winter, giving persons the first-glance impression that the large shrubby cactus, sometimes 8 feet high, is in bloom. The fruits are eaten by cattle. This species is typical of desert grasslands and is most abundant in the open country below the edge of the oak belt in desert mountains. Stems of the dead plants leave a hollow cylinder of attractive wooden meshes when the soft tissues decay, and are favored for making canes, as the stem is long and straight, hence the name walkingstick cholla.

Opuntia spinosior   Cactus Family

61. Evening-primrose

Also called "sun-drops," these plants are particularly welcome because they bloom early in the springtime. Many species of evening-primrose are large flowered, abundant along roadsides and sandy flats, and notably fragrant. White-flowered species are more common, but there are several with yellow flowers. Blossoms open at night and begin to wilt, turning pink during the following day. These are among the handsomest of desert plants and during favorable years make a spectacular spring display, sometimes growing with goldpoppies and sandverbenas to produce a riot of color.

Oenothera trichocalyx   Evening-primrose Family

62. Ocotillo

Common to all of the deserts crossed by the boundary between the United States and Mexico, ocotillo (oh-koh-TEE-yoh) is a spectacular shrub, its many long, stiff, green-barked and thorn-guarded stems bearing at their tips clusters of bright red flowers from April to June. Following rains, the stems cover themselves with clusters of bright green leaves. When drought comes these leaves are shed, to be renewed again after another rain. This procedure may be repeated half a dozen times in one year. Cahuilla Indians eat both flowers and seeds, and make a beverage by soaking the blossoms in water. When planted as hedgerows the thorny wands make an impenetrable fence.

Fouquieria splendens   Ocotillo Family

EUROPEAN GLORYBIND

63. European glorybind

Better known as "bindweed" or "wild morning glory," this naturalized perennial has become a serious agricultural pest throughout the Southwest. In California it is considered the State's worst weed. Once established, its deep root system spreads widely, sending up shoots that grow rapidly with climbing, vine-like stems and morning glory-like white to pink flowers that bloom from May to July. In the desert it is usually found on road shoulders, where it makes an attractive display. The name *convolvulus* comes from the Latin and means "to entwine." A blood-clotting substance has been found in this plant.

Convolvulus arvensis   Convolvulus Family

64. Santa Fe phlox

Usually found in desert mountain ranges, at elevations between 5,000 and 6,000 feet, this ground-hugging, herbaceous perennial blossoms in May and June. Flowers are larger than those of the several other desert species of phlox, most of which have longer flower stems and vary in color from white to purple.

Phlox nana   Phlox Family

SANTA FE PHLOX

65. Starflower

More commonly known as "gilia" in honor of the eighteenth-century Italian botanist Felippo Luigo Gilii, the many species of gilias are common and widespread throughout the deserts of the Southwest at nearly all elevations. Since the flowers are usually small and range in color from white to lavender, pink, and yellow, they are not as well known as more spectacular genera. Some are annuals but there are also many perennial species. Starflower is found from west Texas and Chihuahua to western Arizona at elevations from 1,000 to 8,000 feet on dry plains and mesas, especially on limestone soils. It blossoms from March to October.

Gilia longiflora   Phlox Family

66. Phacelia

Known also as "scorpionweed" and "wild heliotrope," phacelia is a handsome plant with coarse foliage, somewhat hairy and sticky. Among other plants it often grows to a height of 18 inches, but on dry, open desert flats is usually much shorter. Flowers, which may be found from February to June, are sweet scented, but the foliage has a disagreeable odor. *Crenulata,* which is one of many species, grows from New Mexico and southern Utah throughout Arizona to Lower California. It is conspicuous among the springblooming flowers of the desert. The curling flower heads which bear some resemblance to the erect tail of a scorpion are responsible for the name "scorpionweed."

Phacelia crenulata   Waterleaf Family

PURPLEMAT

67. Purplemat

In favorable years these ground-hugging plants form broad, colorful mats, but in dry seasons these annuals may be tiny, each with a single flower almost as large as the rest of the plant. Flowering from February to May, bloom is heaviest in March and April. This species is common on flat, sandy, open desert soils from southeastern California and Baja California to southeastern Arizona at elevations below 3,500 feet. Because of its low-growing habit, purplemat requires that you lie prone to examine it closely, hence is one of the many small desert herbs called "belly-flowers."

Nama demissum   Waterleaf Family

BUFFALOBUR NIGHTSHADE

68. Buffalobur nightshade

Believed to be the original host of the Colorado potato beetle, this annual is a pest on rangelands because of its spine-covered stems and fruits. Spines are long, straight, sharp, and straw-colored. It is common on desert plains and mesas at elevations from 1,000 up to 7,000 feet, blooming from June to August. The leaves and unripe fruits of this and several other species are reportedly poisonous, as they contain an alkaloid, solanin.

Solanum rostratum   Potato Family

69. Purple nightshade

A showy plant when blossoming, purple nightshade attracts attention along roadsides throughout the summer months. It is also common on rocky slopes between 3,500 and 5,500 feet. Plants are from 1 to 3 feet high and grow from a perennial root. Under favorable conditions the flowers are about an inch in diameter and form handsome, loose clusters. The berry-like fruits are pale green to purple and the size of a small cherry. A close relative, *Solanum jamesii,* is known as wild-potato, producing small but edible tubers which are eaten by desert Indians.

Solanum zantii   Potato Family

70. Sacred datura

One of the really striking flowers of the deserts and mesas, the large, showy, trumpet-shaped blossoms and broad, dark green leaves of the datura or "western jimson" arouse the curiosity of persons seeing them for the first time. Quite common along roadsides below 6,000 feet from California to Texas and Mexico, the white blossoms remain open at night but close and droop soon after sunrise. The summer-blooming plants often grow in large clumps with buds, flowers, and maturing fruits all present at the same time. Indians used the plants for various medicinal purposes, a dangerous practice, since all parts of the plant contain various alkaloids, including atropine. Roots are narcotic and were sometimes eaten by Indians to induce visions.

Datura meteloides   Potato Family

TREE TOBACCO

71. Tree tobacco

Sometimes growing to a height of 10 or 12 feet, the graceful swaying branches of tree tobacco bear at their ends clusters of tubular, greenish-yellow blossoms 2 to 3 inches long. The leaves contain the alkaloid anabasine, which is poisonous to livestock. Leaves of the closely related and much smaller desert tobacco, *Nicotiana trigonophylla,* contain nicotine and have long been smoked by desert Indians. The plant is still so used on ceremonial occasions. *Nicotiana* was named for Jean Nicot, French ambassador to Portugal, who introduced tobacco to France about 1560.

Nicotiana glauca   Potato Family

72. Texas silverleaf

Although restricted in its range to the Chihuahuan Desert, silverleaf, or *ceniza* as it is locally known, is so spectacular when in blossom that it invariably attracts attention and arouses interest. The small, abundant, ash-gray leaves give this 3- to 4-foot shrub a distinguished appearance throughout the year, but when it suddenly bursts into bloom, usually in September, it becomes a thing of rare beauty. It is so sensitive to moisture that it may blossom a few hours after a soaking rain, which gives rise to the popular belief that it can forecast wet weather and in consequence it is sometimes called "barometer bush."

Leucophyllum frutescens   Figwort Family

TEXAS SILVERLEAF

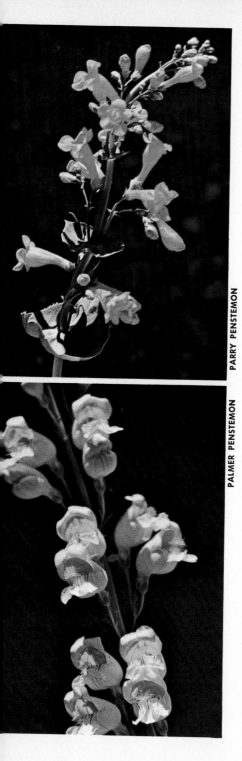

PARRY PENSTEMON

PALMER PENSTEMON

73. Parry penstemon

Penstemons or "beardtongues" of various species are numerous in the desert as well as throughout the higher, moister portions of the Southwest. Parry penstemon is one of the more noticeable desert species because of its showy springtime flowers covering the many erect stems, which are sometimes as much as 4 feet in length. It is fairly common throughout southern Arizona and Sonora at elevations between 1,500 and 5,000 feet, on mesa slopes and mountain canyons. Individuals are usually well scattered, so do not contribute to the mass flower displays of desert springtime.

Penstemon parryi   Figwort Family

74. Palmer penstemon

Known in southern California as "scented penstemon" because of its fragrance, this regal "beardtongue" comes to the height of bloom in May. However, it may be found in flower from March to September. When the tall, flower-covered stems grow in abundance, as often occurs in gravelly washes at elevations between 3,500 and 6,500 feet, the sight is remarkable. This species prefers limestone soils in both the Mojave-Colorado and Sonoran Deserts. The sweet nectar attracts bees.

Penstemon palmeri   Figwort Family

NORTHWESTERN PAINTEDCUP

75. Northwestern paintedcup

Paintedcups, or "Indian paintbrushes" as they are more widely known, are found from desert lowlands to snow-capped mountain tops. *Castilleja linariaefolia* is the State flower of Wyoming. The northwestern paintedcup, known in southern California as "desert paintbrush," has an extremely wide range. The flash of red among other desert plants is actually due to the brightly colored floral bracts, as the flowers themselves are small and inconspicuous. This species blossoms in early spring in rocky or gravelly locations between 2,000 and 7,000 feet, on dry plains and hillsides.

Castilleja angustifolia   Figwort Family

ESCOBITA OWLCLOVER

76. Escobita owlclover

Owlclover is one of the short-stemmed desert spring annuals which, in favorable seasons, carpet the desert floor with a beautiful, colorful mass display. Sometimes growing in pure stands, at others mixed with goldpoppies, lupines, or other spring flowers, it is found throughout southern Arizona, southern California, and Baja California, at elevations between 1,500 and 4,500 feet, blossoming from March to May. Cattle and sheep graze it extensively. The Spanish name *escobita* means "little broom." Individual flowers are not conspicuous, but their clusters intermixed with the colorful bracts produce a pretty, feathery effect.

Orthocarpus purpurascens   Figwort Family

DESERTWILLOW

FLORIDA YELLOWTRUMPET

77. Desertwillow

More properly called "desert catalpa," this tall shrub or small tree, 6 to 15 feet high, has willow-like leaves, spreading branches, and a short, crooked, black-barked trunk. The violet-scented flowers usually appear from April to August, often after the start of summer rains. They are replaced by long, slender seed pods that remain dangling from the branches for months. Mexicans make from the dried flowers a tea that they believe has considerable medicinal value. Desertwillow is usually found along desert washes below 4,000 feet from west Texas to southern California and northern Mexico. It is frequently cultivated as an ornamental because of its attractive orchid-like flowers.

Chilopsis linearis   Bignonia Family

78. Florida yellowtrumpet

A glossy-leafed shrub with golden, trumpet-shaped flowers, the yellowtrumpet bush blooms from May to October on dry, rocky hillsides between elevations of 3,000 and 5,000 feet. It is not common, but occurs from western Texas through southern New Mexico and Arizona southward into tropical America. Yellowtrumpet is cultivated as an ornamental in southern parts of the United States and in Mexico. The roots are used medicinally and in making a beverage. Stems and leaves contain small quantities of rubber. The shrubs, which occasionally reach a height of 6 feet, are browsed by bighorn sheep and probably by deer.

Stenolobium stans   Bignonia Family

79. Louisiana broomrape

Lacking chlorophyll and parasitic on the roots of bur-sage and other desert composites, broomrape is so unusual in appearance as to attract immediate attention. Although fairly common in low-elevation deserts from west Texas and Mexico to southern California, it is occasionally found as far north as southern Utah and Nevada and at elevations up to 7,-000 feet. The rather inconspicuous flowers appear from February to September. Navajo Indians made a decoction of the plant as a treatment for sores. Desert Indians ate the tender stems in springtime.

Orobanche ludoviciana   Broomrape Family

LOUISIANA BROOMRAPE

80. Buffalogourd

Not limited to deserts, this rank-growing, ill-smelling, vine-like plant may have stems up to 20 feet in length. Its large, gray-green leaves and yellow, squash-blossom flowers are conspicuous along roadsides throughout the Southwest. The globular fruits, about the size of tennis balls, were cooked by desert Indians, or dried for winter consumption. The seeds were boiled to form a pasty mush. Because the fruits are eaten by desert animals, they were known in some localities as "coyote melons," and are called *calabazillas* in Spanish-speaking communities. Early Californians used the crushed roots as a cleansing agent in washing clothes, but found particles clinging to the cloth were a skin irritant.

Cucurbita foetidissima   Gourd Family

BUFFALOGOURD

81. Sticky snakeweed

Common throughout the Southwest, particularly on overgrazed rangelands and deserted clearings, this plant, also called "matchweed" or "turpentine-weed," often occurs in almost pure stands. The resinous stems burn readily, throwing off black smoke. Most abundant on dry hills and mesas, 3,000 to 6,000 feet elevation, this perennial is found from 1,000 to 7,000 feet, blossoming from June to October. Bees obtain nectar and pollen from the small but densely crowded, yellow flower clusters. The many stiff, upright branches cause some plants to appear almost globular in shape and a foot to 2 feet in diameter. Plants of this genus are reported as poisonous to sheep and goats if eaten in quantity, but are apparently unpalatable, as they are rarely grazed.

Gutierrezia lucida   Sunflower Family

82. Desertstar

Also known as "desert daisy" and "rock daisy," this dwarf winter annual grows on sandy or stony mesas at elevations below 3,500 feet, blossoming from February through April. The short stems spread to form a mat or rosette, 5 or 6 inches across, growing flat on the sand, and ornamented with many small flowers, each set off by a small cluster of leaves. Desertstar grows principally in southern Arizona and southern California, but has been recorded from southern Utah and Sonora, Mexico.

Monoptilon bellioides   Sunflower Family

83. Mohave aster

Varying in color from violet and lavender to almost white, flower heads of the Mohave aster are numerous, sometimes as many as 20 simultaneously in bloom on one plant. This ornamental perennial prefers dry, rocky slopes below 6,000 feet in southern Utah, Nevada, western Arizona, and southeastern California. Characterized by silvery foliage and large flower heads, the Mohave aster is well worthy of cultivation and does well in hot, dry locations. Flowers appear from March to May, but with the coming of summer heat the stems and leaves become twisted, brown, and unattractive.

Aster abatus   Sunflower Family

84. Spreading fleabane

By no means limited to the deserts, fleabane is common throughout the Southwest, including parts of Mexico. In some localities it is known as "wild-daisy." Six to 15 inches tall, with attractive circular flowers, fleabane often forms noticeable patches along road shoulders and on dry open slopes, blossoming from February to October. Flowers may be an inch in diameter in springtime, but those in summer are usually smaller. The name arises from an ancient belief that the odor of some species repelled fleas.

Erigeron divergens   Sunflower Family

MOHAVE ASTER

SPREADING FLEABANE

85. Broom baccharis

Locally called "desert-broom," or "Mexican broom," this species of baccharis is an erect, coarse, evergreen shrub 3 to 6 feet high, frequently encountered on hillsides and bottomlands at elevations between 1,000 and 5,500 feet from southwestern New Mexico to southern and Baja California and northern Mexico. Greening up following summer rains, the shrubs blossom from September to February. Flowers are inconspicuous, but the fruits develop as masses of spectacular cottony threads, giving the shrubs a snow-covered appearance. Among some Indian tribes the twigs are chewed to relieve toothache. In Mexico the shrub is called *hierba del pasmo*.

Baccharis sarothroides   Sunflower Family

BROOM BACCHARIS

86. Desert zinnia

From 3 inches to a foot high, desert zinnia is a dwarf shrub with small, stiff, dull green leaves and attractive, four-petaled flowers that are present from April to October. Preferring clayey or rocky, arid soils at elevations 2,500 to 5,000 feet, this species is found from west Texas to southern Arizona and Mexico. Although related to the garden zinnia, which is a native of Mexico, only the large flowered desert species, *Zinnia grandiflora,* is considered worthy of cultivation.

Zinnia pumila   Sunflower Family

DESERT ZINNIA

87. White brittlebush

Sometimes blossoming as early as November and often lingering until May, brittlebush is a dome-shaped, winter-flowering bush that brings delight to desert dwellers in Nevada, Arizona, southern California, and northwestern Mexico. Stems of the low-growing, silvery-leaved shrub exude a gum which was chewed by desert Indians and burned as incense by priests in mission churches, giving the plant the local name, *incienso*. Strictly a desert shrub, about 3 feet high, brittlebush prefers rocky hillsides below 3,000 feet. Growing in masses it often covers entire slopes with a mass of golden bloom, contributing to the early spring flower display. Bighorn sheep are reported to rely on this species for browse.

Encelia farinosa   Sunflower Family

WHITE BRITTLEBUSH

SILVERLEAF ENCELIOPSIS

88. Silverleaf enceliopsis

Restricted in its range to the region in which Utah, Arizona, and Nevada meet, the "giant sunray," as it is sometimes called, is spectacular rather than beautiful. Coarse and weedy, the large clusters of silvery leaves and long stemmed, sunflower-like blossoms that appear from April to June invariably attract attention and stimulate curiosity. An even larger species, *Enceliopsis covillei,* with blossoms up to 6 inches in diameter, is found in canyons on the west side of the Panamint Mountains in California.

Enceliopsis argophylla   Sunflower Family

GOLDEN CROWNBEARD

DOUGLAS COREOPSIS

89. Golden crownbeard

Although it is reported from elevations up to 7,000 feet, golden crownbeard is usually found at much lower levels from Kansas south to Texas, California, and northern Mexico. Sometimes growing in clusters, single plants are also common as a weed of roadsides and waste ground. The all-yellow, sunflower-like blossoms are widespread in the desert from April to November. Desert Indians and early pioneers are said to have used the plant to treat boils and skin diseases. The Hopis soaked the plants in water in which they bathed, to relieve the pain of insect bites.

Verbesina encelioides   Sunflower Family

90. Douglas coreopsis

Also called "tickseed," wild coreopsis is closely related to cultivated ornamentals of the same name. The desert species inhabits open locations at elevations between 1,500 and 2,500 feet in southern Arizona, southern California, and Baja California. Plants usually bloom between February and May. The closely related *Coreopsis bigelovii* is a southern California annual having somewhat larger flowers, up to 2 inches in diameter, with orange centers. Flower stems are naked, with the leaves clustered at their bases.

Coreopsis douglasii   Sunflower Family

91. Whitestem paperflower

At its best in sandy desert soil, paperflower is a compact, shrubby plant about 1 foot high, with tangled branches. When fully developed it is symmetrically globular in outline. It prefers mesas and desert plains at elevations between 2,000 and 5,000 feet from western New Mexico to southern California and northern Mexico, flowering throughout the year but most abundantly in springtime. Sometimes called "paper-daisy," the flowers are persistent, fading to straw color and turning papery with age. They may remain on the stems for weeks.

Psilostrophe cooperi   Sunflower Family

WHITESTEM PAPERFLOWER

DESERT BAILEYA

92. Desert baileya

Commonly called "desert marigold," baileya blossoms in all seasons, most heavily from March to November, and is one of the better known flowers of the Southwest. Each circular blossom occupies the tip of a foot-high stem. Plants usually have a thrifty, garden-variety appearance. They are common along roadsides and on well-drained, gravelly slopes up to 5,000 feet from west Texas to southeastern California and Chihuahua. The large flower heads are showy and the species is cultivated in California. Cases are on record of sheep and goats on overgrazed ranges being poisoned by eating this plant.

Baileya multiradiata   Sunflower Family

93. Branchy goldfields

Covering vast stretches of open desert with a carpet of yellow bloom following wet winters, goldfields is an appropriately named spring flower found at elevations below 4,500 feet. The low-growing plant produces small but attractive blossoms on mesas and plains, March to May, from central and southern Arizona to California, and Baja California. Horses graze *Baeria* avidly, but are annoyed by a small fly that frequents the fragrant blossoms, giving the plant the name "fly flower" in some localities.

Baeria crysostoma   Sunflower Family

94. Chaenactis

Probably because it is one of the attractive white desert flowers, chaenactis is popularly called "morning bride." A larger, yellow-flowered species, *Chaenactis lanosa,* found on the California deserts, is called "golden girls." Both are spring flowering annuals and, in common with other members of the genus, sometimes called "pincushion plants." "Morning bride" is often found growing about the bases of creosotebushes, thriving at elevations between 1,000 and 3,500 feet in southern Nevada, western Arizona, and southeastern California.

Chaenactis fremontii   Sunflower Family

95. Douglas groundsel

Rarely considered beautiful, the groundsels are common and widespread, and are readily recognized by the untidy appearance of the large flowers which are sometimes almost 2 inches in diameter. The rather delicate, stringy foliage is sometimes covered with cottony threads. One species is called "ragwort." Douglas groundsel is a shrubby plant sometimes as much as 3 feet high, common in sandy washes and on dry foothill slopes. It occurs from southern Utah and Arizona to California and Mexico, between 1,000 and 6,000 feet. At lower elevations these plants bloom at almost any time of year.

Senecio douglasii   Sunflower Family

DOUGLAS GROUNDSEL

NEW MEXICO THISTLE

96. New Mexico thistle

Everyone recognizes the thistles with their prickly leaves and stems, and large flowers ranging in color from white to lavender, pink and purple. Several species grow in the deserts, the New Mexico species being widespread at elevations from 1,000 to 6,000 feet in Colorado and Nevada south through New Mexico and Arizona to California, blossoming from March to September. Navajo and Hopi Indians are reported to use thistles medicinally. The nectar of some species is eagerly sought by hummingbirds.

Cirsium neomexicanum   Sunflower Family

97. Desert dandelion

A very attractive plant, desert dandelion has several flower stalks from a few inches to a foot tall. Some of the blossoms may be nearly 2 inches in diameter. This annual is common in open, sandy basins, where it is a conspicuous contributor to the spring flower spread, blooming from March through May in the creosotebush belt of Arizona and southern California. It has been reported from as far north as Idaho and Oregon. Sometimes a single plant has 10 or 12 flower heads in blossom at the same time.

Malacothryx glabrata   Sunflower Family

98. Malacothryx

There are many species of malacothryx native to the western and southwestern United States. Some are locally called "desert dandelion," "snake's head," "yellow saucers," and "cliff aster." *Fendleri* is one of the smaller species, with stems only 4 or 5 inches long, rising from a rosette of bluish-green leaves. Blooming from March to June, this delicate relative of the common dandelion covers with its pale yellow flowers rocky slopes and sandy plains and mesas, at elevations between 2,000 and 5,000 feet from West Texas to western Arizona.

Malacothryx fendleri   Sunflower Family

99. Tackstem

Called tackstem because of the numerous dark-colored, tack-shaped glands protruding from the stem, this white-flowered, branching annual blossoms from March to May at elevations of 500 to 4,000 feet. It is a conspicuous item of the spring flower display from west Texas to southern California and northern Mexico. A similar species with yellow flowers, *Calycoseris parryi*, common at elevations around 3,000 feet, blooms in March and April. It is found in southwestern Utah, Arizona, and southern California.

Calycoseris wrightii   Sunflower Family

TACKSTEM

PRICKLY SOWTHISTLE

100. Prickly sowthistle

Naturalized from Europe and generally considered a weed, sowthistle is found in waste grounds and along roadsides from near sea level to 8,000 feet. It blossoms from February to August, the flowers becoming cottony seed heads as conspicuous as the blooms. A close relative, *Sonchus oleraceus*, which blossoms from March to September, produces a gum from the drying of the sap, reportedly a powerful cathartic. It has also been used as a treatment for persons suffering from the habitual use of opium derivatives.

Sonchus asper   Sunflower Family

Suggestions for Additional Reading

Armstrong, Margaret, *Field Book of Western Wild Flowers,* C. P. Putnam's Sons, New York, 1915.

Benson, Lyman, *The Cacti of Arizona,* University of Arizona Press, Tucson, 1950.

Benson, Lyman, and Darrow, Robert, *The Trees and Shrubs of the Southwestern Deserts,* University of New Mexico Press, Albuquerque, N.M., 1954.

Dodge, Natt, *Flowers of the Southwest Desert,* Southwestern Monuments Association, Globe, Arizona, 1951.

Hornaday, W. T., *Camp-fires on Desert and Lava,* Charles Scribner's Sons, New York, 1909.

Jaeger, Edmund C., *Desert Wild Flowers,* Stanford University Press, Stanford, California, 1956.

Jaeger, Edmund C., *The North American Deserts,* Stanford University Press, Stanford, California, 1957.

Lemmon, Robert S., and Johnson, Charles C., *Wildflowers of North America in Full Color,* Hanover House, Garden City, N.Y., 1961.

Leopold, A. Starker, *The Desert,* (Life Nature Library) Time Inc., New York, 1961.

McDougall, W. B., and Sperry, Omer E., *Plants of Big Bend National Park,* Government Printing Office, Washington, D. C., 1951.

Shreve, Forrest, and Wiggins, Ira L., *Vegetation and Flora of the Sonora Desert,* Carnegie Institution of Washington Publication No. 591, Vol. 1, Washington, D. C., 1951.

Vines, Robert A., *Trees, Shrubs, and Woody Vines of the Southwest,* University of Texas Press, Austin, 1960.

Index

Agave	*Agave scabra*	14
Arizona jumping cholla	*Opuntia bigelovii*	57
Beavertail pricklypear	*Opuntia basilaris*	55
Bladderpod	*Lesquerella gordonii*	24
Blue paloverde	*Cercidium floridum*	31
Branchy goldfields	*Baeria crysostoma*	93
Broom baccharis	*Baccharis sarothroides*	85
Buffalobur nightshade	*Solanum rostratum*	68
Buffalogourd	*Cucurbita foetidissima*	80
Caltrop	*Kallstroemia grandiflora*	41
Canaigre	*Rumex hymenosepalus*	17
Carneros yucca	*Yucca carnerosana*	11
Catclaw acacia	*Acacia greggii*	27
Chaenactis	*Chaenactis fremontii*	94
Claretcup echinocereus	*Echinocereus triglochidiatus*	49
Coast erysimum	*Erysimum capitatum*	25
Common reed	*Phragmites communis*	2
Coulter lupine	*Lupinus sparsiflorus*	33
Coville creosotebush	*Larrea tridentata*	40
Crested pricklepoppy	*Argemone platyceras*	21
Deerhorn cactus	*Peniocereus greggii*	46
Desert baileya	*Baileya multiradiata*	92
Desert bearpoppy	*Arctomecon meriamii*	22
Douglas coreopsis	*Coreopsis douglasii*	90
Desert dandelion	*Malacothryx glabrata*	97
Desert globemallow	*Sphaeralcea ambigua*	42
Desertlily	*Hesperocallis undulata*	4
Desert mariposa	*Calochortus kennedyi*	7
Desert sandverbena	*Abronia villosa*	19
Desertstar	*Monoptilon bellioides*	82
Desertwillow	*Chilopsis linearis*	77
Desert zinnia	*Zinnia pumila*	86
Douglas groundsel	*Senecio douglasii*	95
Engelmann echinocereus	*Echinocereus engelmannii*	50
Engelmann pricklypear	*Opuntia engelmannii*	56
Escobita owlclover	*Orthocarpus purpurascens*	76
European glorybind	*Convolvulus arvensis*	63
Evening-primrose	*Oenothera trichocalyx*	61
False-mesquite calliandra	*Calliandra eriophylla*	26
Fishhook mammillaria	*Mammillaria microcarpa*	54
Five-stamen tamarisk	*Tamarix pentandra*	43
Florida yellowtrumpet	*Stenolobium stans*	78
Fremont dalea	*Dalea fremontii*	36
Golden crownbeard	*Verbesina encelioides*	89
Golden segolily mariposa	*Calochortus nuttalii*	6

Honey mesquite	*Prosopis juliflora*	29
Inyo lupine	*Lupinus excubitus*	34
Joshuatree yucca	*Yucca brevifolia*	9
Lechuguilla agave	*Agave lophantha poselgeri*	16
Longleaf ephedra	*Ephedra trifurca*	1
Louisiana broomrape	*Orobanche ludoviciana*	79
Malacothryx	*Malacothryx fendleri*	98
Mescat acacia	*Acacia constricta*	28
Mexican goldpoppy	*Eschscholtzia mexicana*	20
Mohave aster	*Aster abatus*	83
New Mexico thistle	*Cirsium neomexicanum*	96
Northwestern paintedcup	*Castilleja angustifolia*	75
Ocotillo	*Fouquieria splendens*	62
Organpipe cactus	*Lemaireocereus thurberi*	48
Palmer penstemon	*Penstemon palmeri*	74
Paradise poinciana	*Poinciana gilliesii*	32
Parry agave	*Agave parryi*	15
Parry penstemon	*Penstemon parryi*	73
Phacelia	*Phacelia crenulata*	66
Prickly sowthistle	*Sonchus asper*	100
Purplemat	*Nama demissum*	67
Purple nightshade	*Solanum zantii*	69
Rainbow echinocereus	*Echinocereus rigidissimus*	51
Sacahuista	*Nolina microcarpa*	12
Sacred datura	*Datura meteloides*	70
Saguaro	*Cereus giganteus*	47
Samija mentzelia	*Mentzelia involucrata*	44
Santa Fe phlox	*Phlox nana*	64
Senna	*Cassia bauhinioides*	30
Silverleaf enceliopsis	*Enceliopsis argophylla*	88
Smokethorn	*Dalea spinosa*	35
Soaptree yucca	*Yucca elata*	8
Southwest barrel-cactus	*Ferocactus wislizenii*	53
Spectaclepod	*Dithyraea wislizenii*	23
Spiderwort	*Tradescantia occidentalis*	3
Spreading fleabane	*Erigeron divergens*	84
Starflower	*Gilia longiflora*	65
Sticky snakeweed	*Gutierrezia lucida*	81
Stingbush	*Eucnide urens*	45
Tackstem	*Calycoseris wrightii*	99
Tesajo	*Opuntia leptocaulis*	58
Tesota	*Olneya tesota*	37
Texas heronbill	*Erodium texanum*	39
Texas silverleaf	*Leucophyllum frutescens*	72

Torrey yucca   *Yucca torreyi* 10
Trailing allionia   *Allionia incarnata* 18
Tree tobacco   *Nicotiana glauca* 71

Walkingstick cholla   *Opuntia spinosior* 60
Weakstem mariposa   *Calochortus flexuosus* 5
Wheeler sotol   *Dasylirion wheeleri* 13
Whipple cholla   *Opuntia whipplei* 59
White brittlebush   *Encelia farinosa* 87
Whitestem paperflower *Psilostrophe cooperi* 91
Woolly loco   *Astragalus mollissimus* 38

Yellow pitaya echinocereus   *Echinocereus dasyacanthus* 52